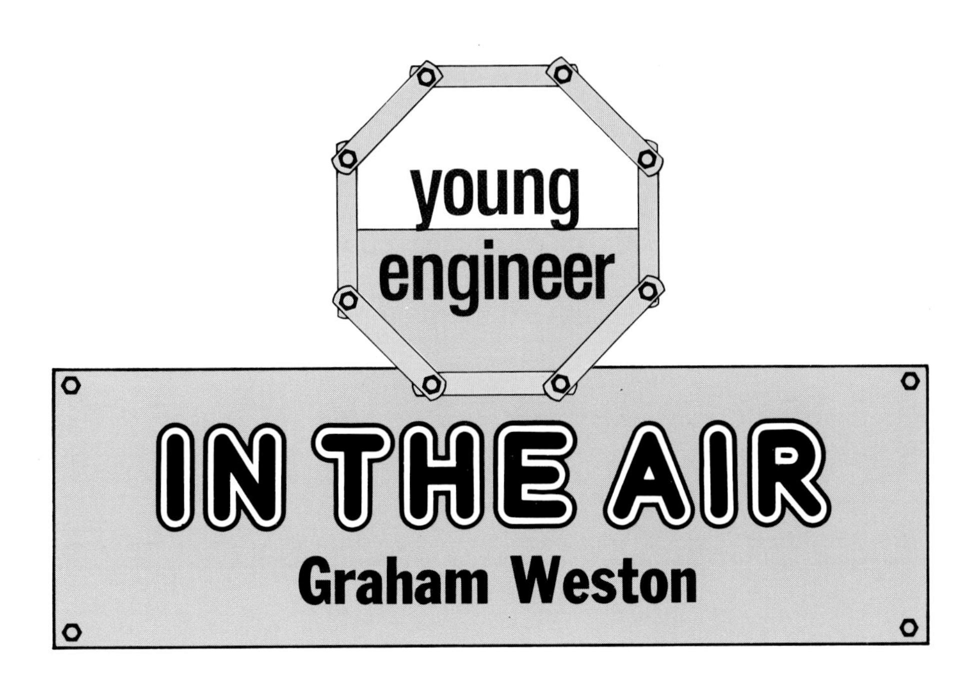

A Young Engineer Book

The Bookwright Press
New York · 1983

In the Air

On the Road

In the Factory

In Communication

First published in the United States in 1983 by
The Bookwright Press, 387 Park Avenue South,
New York, NY 10016
First published in 1983 by
Wayland Publishers Limited, England
ISBN: 0-531-04699-0
Library of Congress Catalog Card Number 83-71636

Illustrated by Gerald Wood
Printed in Italy by G. Canale & C.S.p.A.

Contents

Learning from nature

Birds

Humans have always dreamed of being able to fly. In order to make this dream come true, many lessons had to be learned from nature.

Watch the birds near your home or your school. If you can feed the birds regularly, you will be able to watch them closely.

Draw pictures to show the positions of their wings during flight.

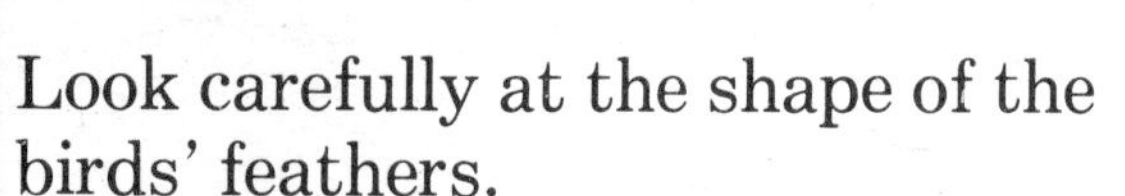

Look carefully at the shape of the birds' feathers.
Are all the feathers the same shape and size?
Collect any feathers you find, and try to discover which birds they came from and from what part of the bird's body.
Look at the feathers under a microscope. Record what you see.

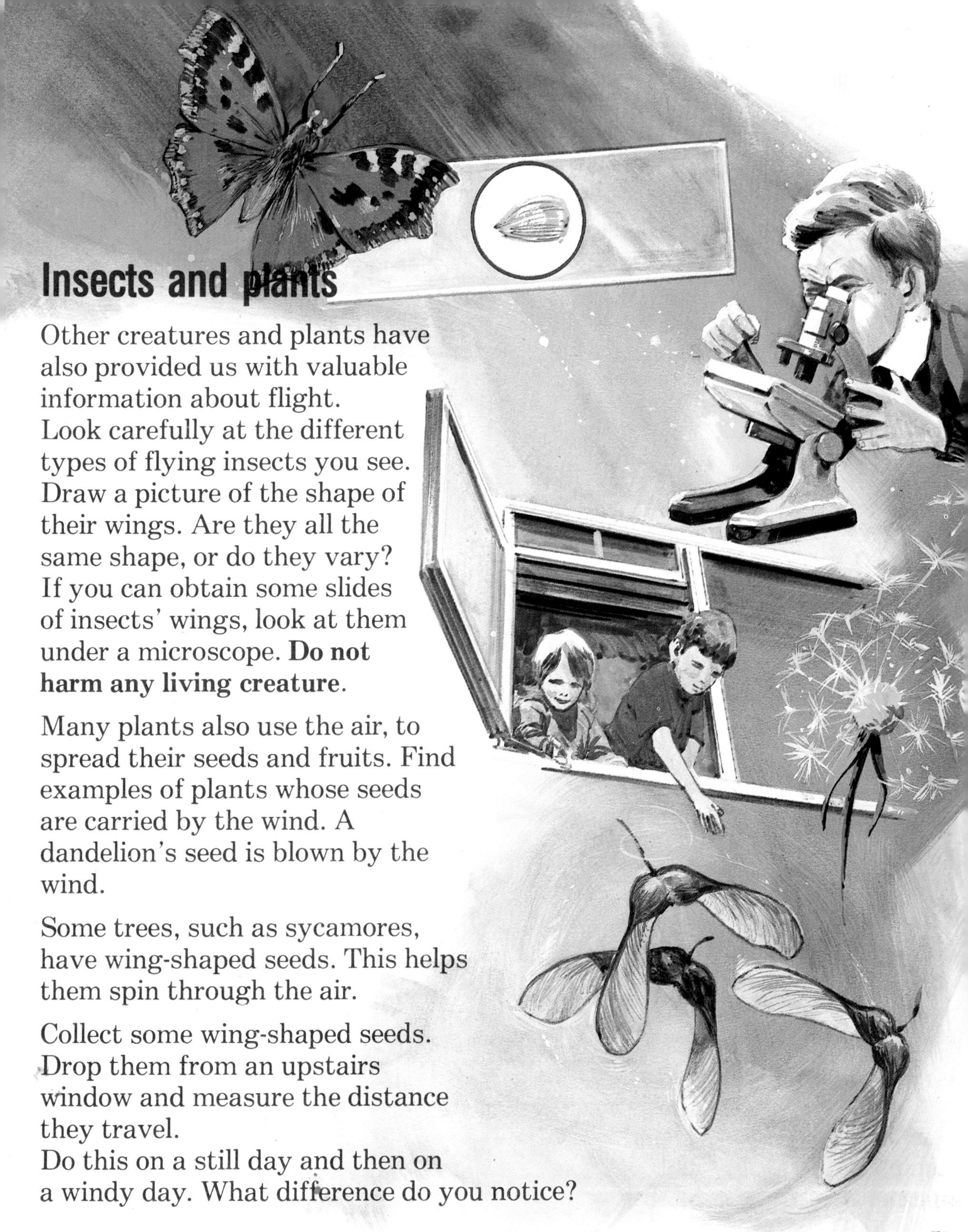

Insects and plants

Other creatures and plants have also provided us with valuable information about flight.
Look carefully at the different types of flying insects you see. Draw a picture of the shape of their wings. Are they all the same shape, or do they vary? If you can obtain some slides of insects' wings, look at them under a microscope. **Do not harm any living creature**.

Many plants also use the air, to spread their seeds and fruits. Find examples of plants whose seeds are carried by the wind. A dandelion's seed is blown by the wind.

Some trees, such as sycamores, have wing-shaped seeds. This helps them spin through the air.

Collect some wing-shaped seeds. Drop them from an upstairs window and measure the distance they travel.
Do this on a still day and then on a windy day. What difference do you notice?

Gravity and parachutes

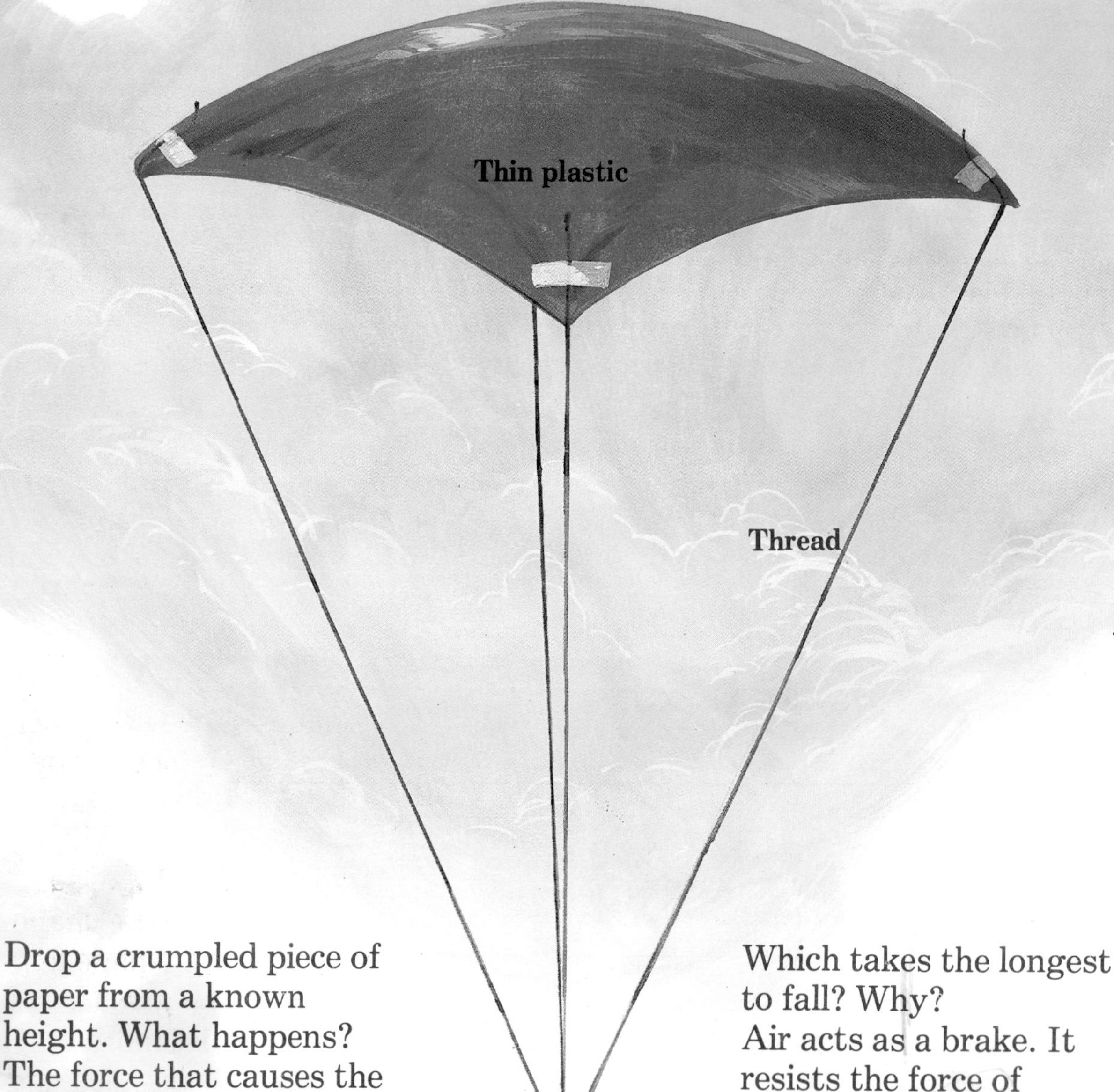

Drop a crumpled piece of paper from a known height. What happens? The force that causes the paper to fall is called gravity.
Now take two similar sheets of paper. Crumple one sheet into a ball. Drop the two pieces of paper from the same height several times.

Which takes the longest to fall? Why?
Air acts as a brake. It resists the force of gravity on things passing through it. People have observed this in nature and used the idea to make parachutes.
Here is a simple parachute for you to make.

Parachute shapes

10 inches
Radius

Make more parachutes from plastic squares of different sizes. Try squares of 8 inches, 16 inches, and 20 inches. Test-fly them. Which one falls slowest? Try adding a variety of weights, and time the descent for each one. Carry out your drop from the same point each time.

Does the weather affect the descent of your parachutes?

Now make a circular parachute.

Draw a circle with a 10-inch radius. Mark off eight equally spaced points on the circumference. Attach equal lengths of strong thread to these points, and tie all eight pieces to a weighted hook. How does your circular parachute compare with the square ones?

Try making parachutes with other shapes. Which shape do you think is the best?

Most modern parachutes have the shape of an airfoil, (page 21). This shape gives a parachutist more control over speed and direction — The parachutist can "fly" the chute. Other shapes merely reduce the speed at which the parachutist falls through the air.

Chapter 2 Balloons

Hot-air balloons

On November 21, 1783, a hot-air balloon built by the Montgolfier brothers made its first manned flight. The hot air inside the balloon was made by burning wood and straw, and the flight lasted 25 minutes.

Nowadays, hot-air balloons can stay up for many hours and travel long distances.

Do you know why hot-air balloons rise into the air? The experiments on pages 9 and 10 will help you to understand.

Experimenting with hot air

Blow up a rubber balloon, but do not blow it up completely. Tie a knot in the neck and measure the circumference of the balloon. Place it on a warm radiator. After 5 minutes, measure it again.
Now place it in a refrigerator, and measure it after 5 minutes.
Does the circumference change?
Can you think why this happens?

Tape an accurate room thermometer to a pole — a broom handle will do. Make sure that the bulb of the thermometer is not touching the pole. Measure the temperature near the floor of a room. Now measure the temperature high up in the room, close to the ceiling. What do you notice about the two readings you have taken? Can you explain this?

These experiments should show you that air takes up more room when it is hot. Hot air is lighter than cold air, and we say that hot air rises.

Upthrust

Find a glass container and half-fill it with water. Take a table-tennis (ping pong) ball and push it into the water until it is just below the surface. Can you feel the ball pressing against your fingers?

This pressure is a result of what is called upthrust. When any object is placed in a liquid or a gas, it will experience an upthrust. If the upthrust is greater than the weight of the object, then the object will rise.

The pressure you felt from the table-tennis ball was caused by the fact that the upthrust was greater than the weight of the ball, and so the ball tried to rise. Your table-tennis ball does not rise up in the air. Although its weight stays the same, the upthrust caused by the air is much less than that caused by the water.

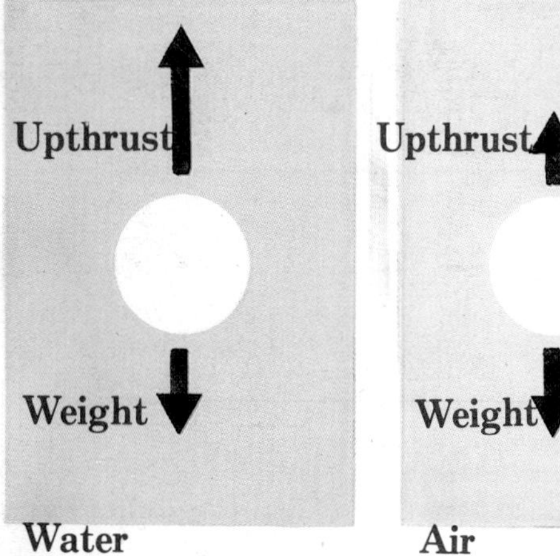

But, as you discovered in the experiments on page 9, hot air is lighter than cold air. So, when the weight of a hot-air balloon is less than the upthrust of the cold air around it, the balloon rises.

Making a hot-air balloon

Take three sheets of tissue paper. Overlap them and glue the overlaps to form one sheet. Do this six times to make six large sheets.
Wait for the glue to dry. Check that there are no holes where the sheets overlap.

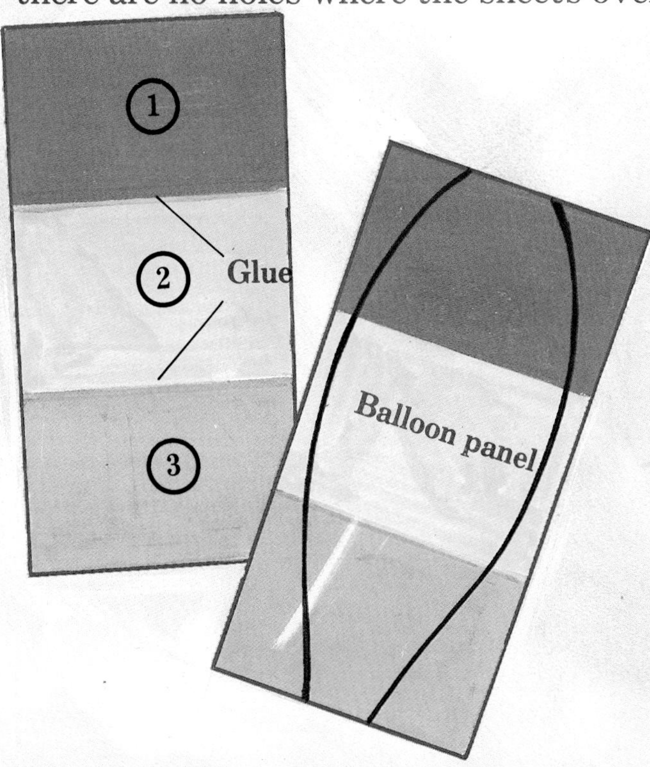

Disk

Collar

Put all six sheets in a pile and carefully cut out the shape of a balloon panel.

Glue the six panels together to make a balloon. Make sure that there are no leaks between the panels. Strengthen the top of the balloon with a tissue paper disk.

Fold another sheet of tissue paper in four to make a thick strip. Glue this strip at the bottom of the balloon to form a collar.

Use warm air from an electric convector heater or a hairdryer to inflate your balloon. Does it fly?

Helium-filled balloons

Helium gas is often used in balloons because it is lighter than air. You can sometimes find helium-filled balloons at fairs or amusement parks. Buy one and try this experiment.

Go outside on a still day. Take a piece of string 30 feet long, and tie one end to a helium-filled balloon. Tie the other end to a heavy object, such as a brick. Place the brick on the ground and hold the balloon close to it. Let go of the balloon and, with a stopwatch, record the time it takes to rise.

Tie a paperclip hook to your balloon and hang washers from it. Time your balloon with one washer, then with two, and so on. What do you notice about the times? Find out how balloonists control the speed at which their balloons rise and fall.

What other gases have been used to fill balloons? How well did they work?

Chapter 3 Gliders

A paper glider

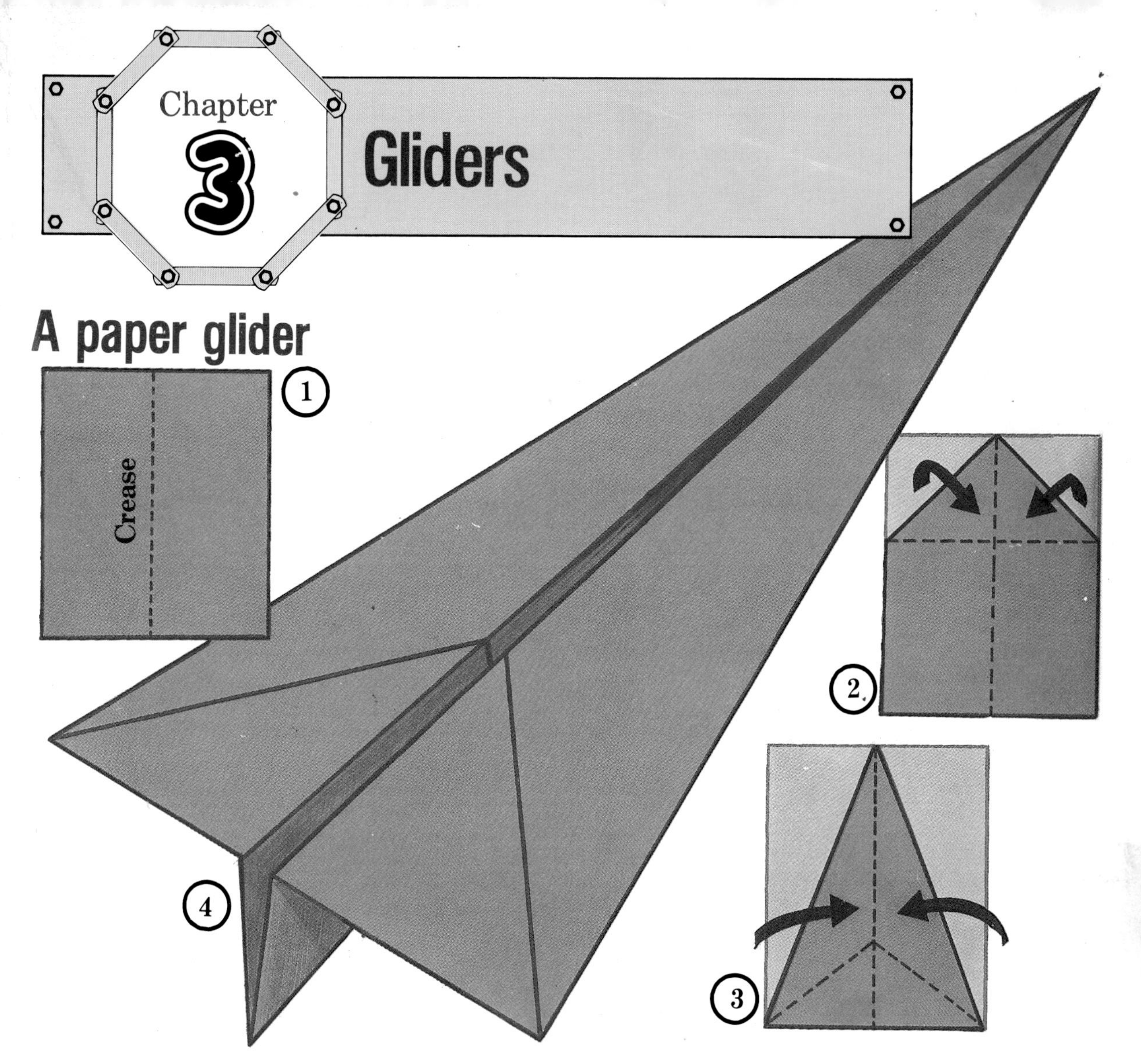

Cut a rectangular piece of paper, about 12 inches by 8 inches.
Make a crease down the center (Fig. 1).
Fold the top two corners down into the middle so that they just touch (Fig. 2).
Fold the two triangles inward again so that they touch the center crease (Fig. 3).
Raise the sides by folding the crease again.
Now fold the wings down at right angles to the body, so that the center crease becomes a keel to hold your glider by (Fig. 4).
Launch your glider and see how it flies.
Where does it fly best?
Does it fly better with the wind or into the wind?
How far will it fly?

Building a balsa wood glider

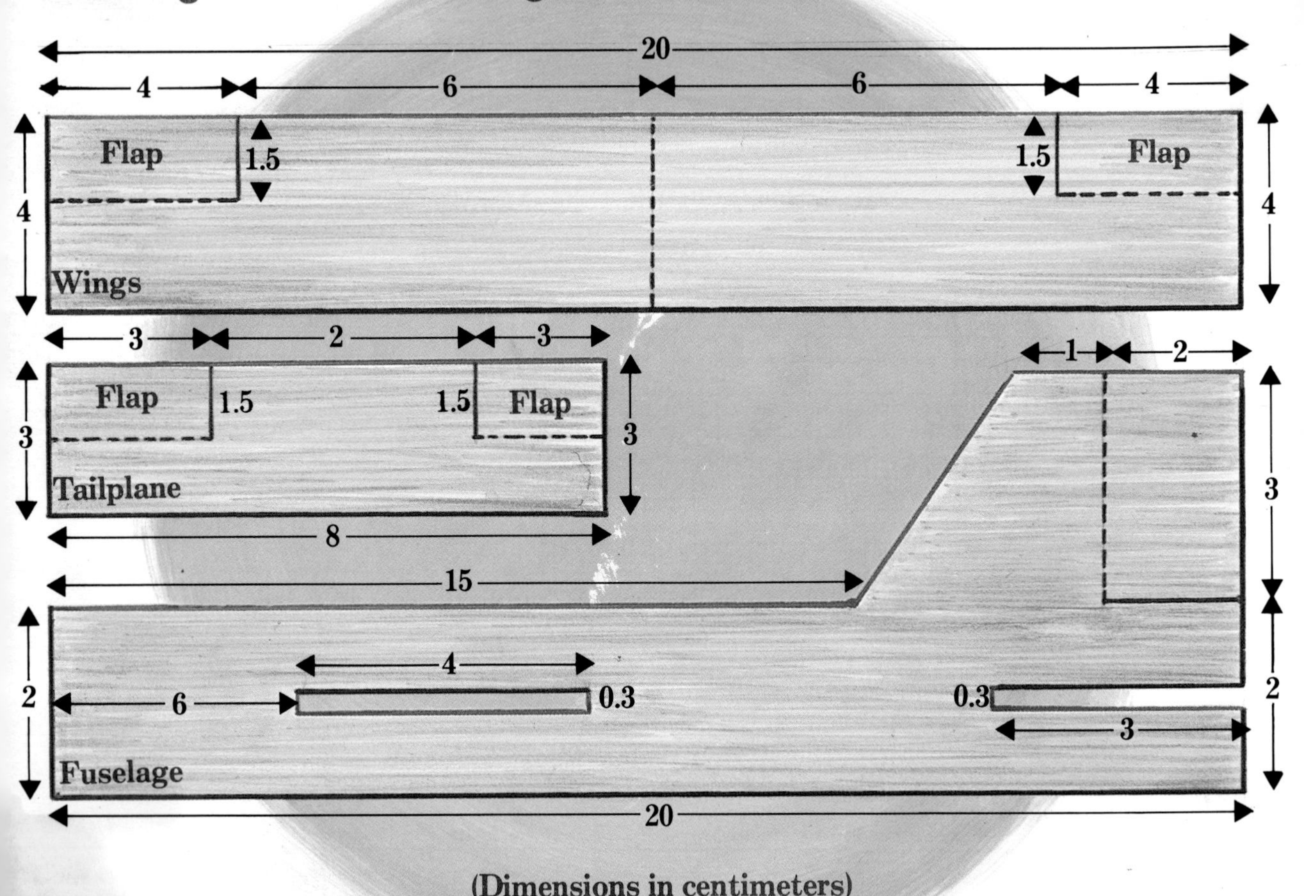

(Dimensions in centimeters)

Buy a sheet of balsa wood 1/8 inch (3mm) thick, and mark out the shapes in the diagram. (The dimensions are all in centimeters.) Then cut along the solid black lines. Do not cut along the dotted lines; simply score them lightly so that they will bend rather than break.

Slide the tailplane into the slot at the back of the fuselage and glue it into position. Make sure that the movable flaps are at the rear. Slide the wings into the slot in the middle of the fuselage so that the flaps are toward the rear. Bend wings upward slightly and then glue them into position. You can use pins to hold them in place while the glue dries. Using a protractor, make sure that both wings are bent upward at the same angle.

When the glue is dry, test-fly your glider. If necessary, add some modeling clay to the front of the glider to balance it.

Bend the rudder to the left and throw your glider. What happens?

Now bend the rudder to the right. What do you think will happen? Throw your glider. Were you right?

The rudder

A glider handles in much the same way as a powered airplane. We will now look at the controls on your glider. The movable part on the tail fin is called the rudder.

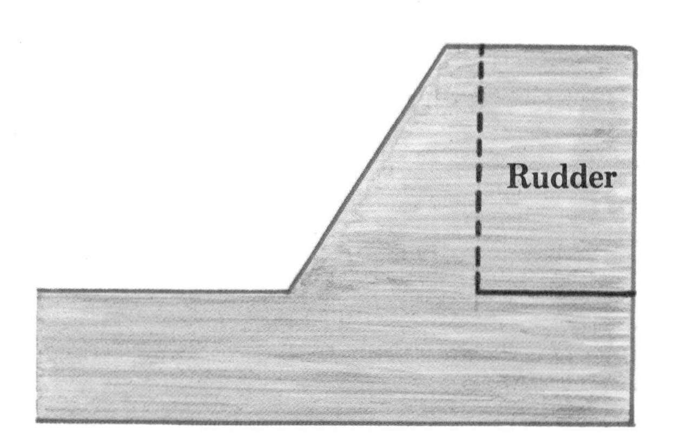

The rudder is not the only control which is used to steer a real airplane — the flaps on the wings are also very important.

The ailerons

The two movable flaps on the wings are called the ailerons.

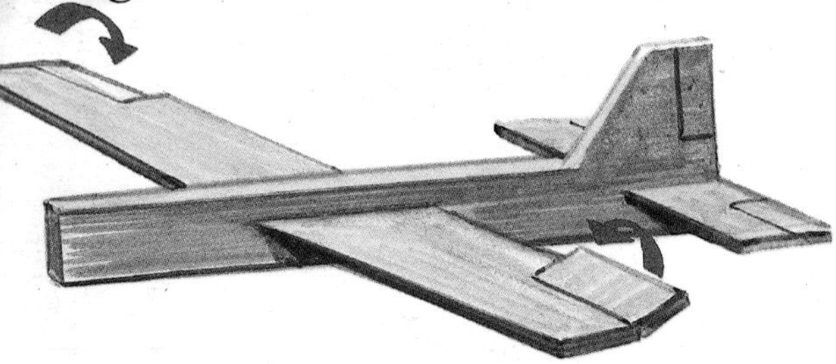

Make sure the rudder is straight. Now bend the right aileron down and left aileron up. Throw the glider. What happens?

Bend the left aileron down and the right aileron up. What happens?

Ailerons are normally used in this way — one up and one down. This has the effect of making the plane roll. It gives the pilot sideways control. What happens if you use only one aileron, or move them both in the same direction?

Bend one aileron up and the other down, and move the rudder to one side. Throw the glider. Does it turn smoothly? Try moving the rudder the other way. What happens?

The elevators

On your tail plane there are two movable parts, called elevators.

Set the rudder and the ailerons straight, and push both elevators into an upward position. What happens when you throw your glider? Now turn both elevators down. What happens? The elevators control the up and down movements of an airplane. When the elevators are raised, the plane will climb. When they are lowered, the plane dives.

Now try adjusting all three sets of controls — the rudder, the ailerons and the elevators. How do you have to move them to make your glider fly best?

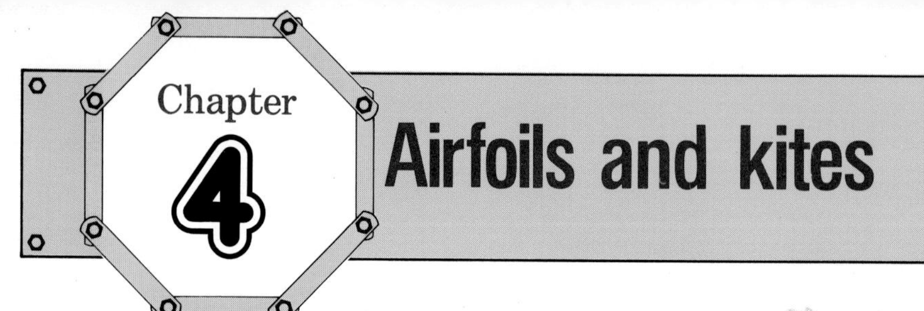

Chapter 4 Airfoils and kites

Moving air

These experiments will help you to see what happens when objects move through the air.

Suspend two table-tennis balls on equal lengths of thread. Fix them with tape. Hang the balls so that they are about 1 inch apart.
With a straw, blow gently between the table-tennis balls to attempt to move them apart. Which way do the balls move? Try blowing harder. What happens?

Cut an oblong, 8 inches by 2½ inches from stiff paper.
Fold down 1¼ inches at each end to make a bridge.
Place your paper bridge on a flat surface.
Blow steadily under the bridge.
What happens to the top of the bridge?
What happens to the sides of the bridge?

Air and lift

Hold a sheet of paper close to your mouth. Hold it so that it hangs down in a curve. Blow across the top surface of the paper. What happens to the paper?

Cut a piece of light cardboard 2 inches square. Draw lines from corner to corner to find the mid-point. Push a thumbtack through the mid-point. Place a cotton thread spool on the card so that the hole in the spool is over the point of the thumbtack.

Hold the cardboard tightly against the spool while you blow down the center. As you blow, let go of the cardboard. It should not drop. The harder you blow the more firmly the cardboard should be held.

This happens because when you blow, the air accelerates through the narrow gap between the cardboard and the spool. The air pressure here is lower than it is below the cardboard and so the cardboard lifts.

Making an airfoil

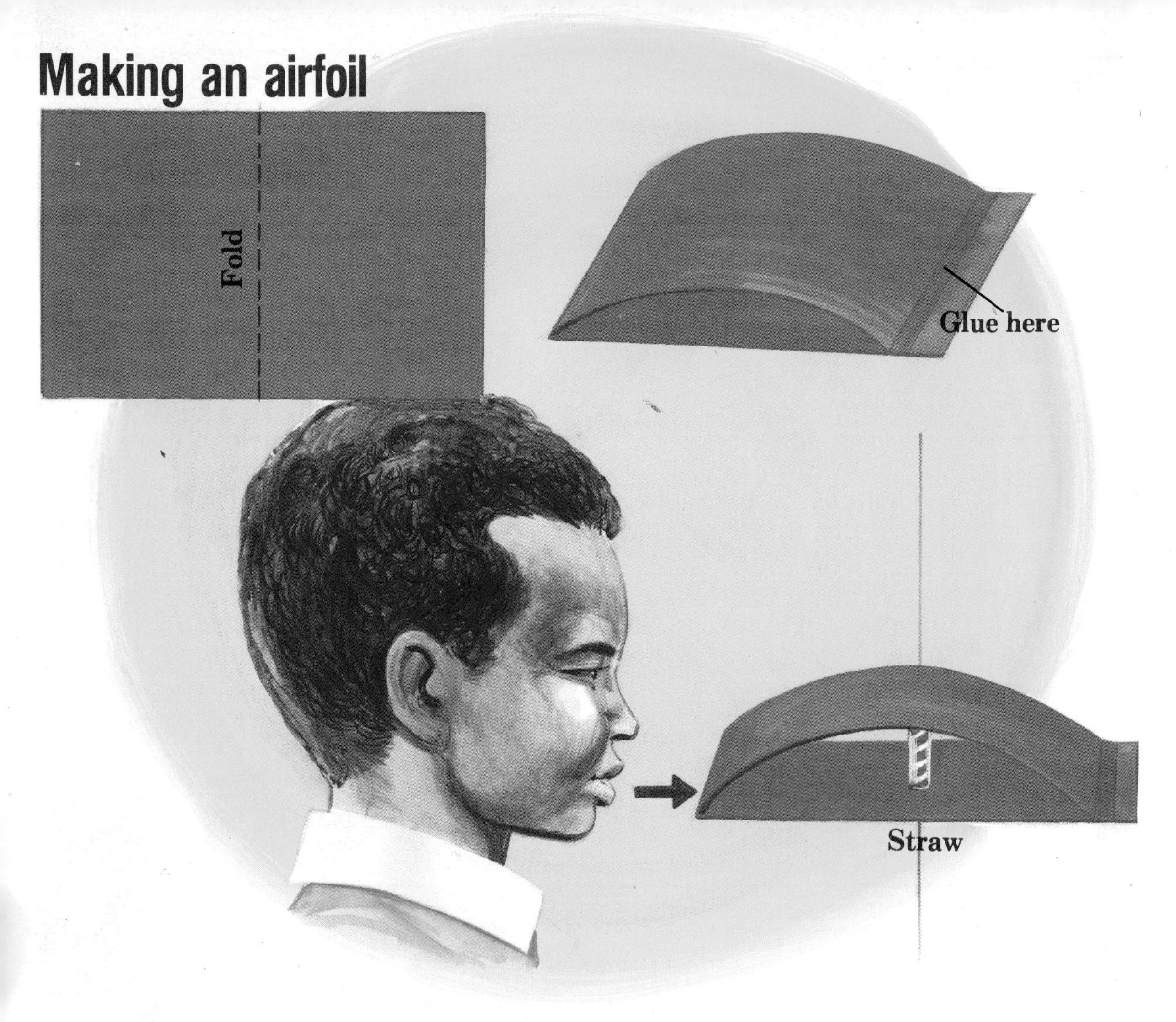

Take a sheet of paper and fold it in half.
Crease the fold firmly.
Bend the top part so that it forms a curve and then guide it in place.
Make a small hole in the top part, and another one below it in the bottom part.
Cut a short length from a straw.
Run some thread through the hole in the top of your airfoil, then through the straw and the hole in the bottom.
The straw will keep your airfoil from collapsing. Hold both ends of the thread.
Hole the thread vertically and blow toward the creased end of the airfoil. What happens?
Try blowing at the other end of the airfoil. What happens?
What happens if you turn the airfoil upside down and blow?

The airfoil shape

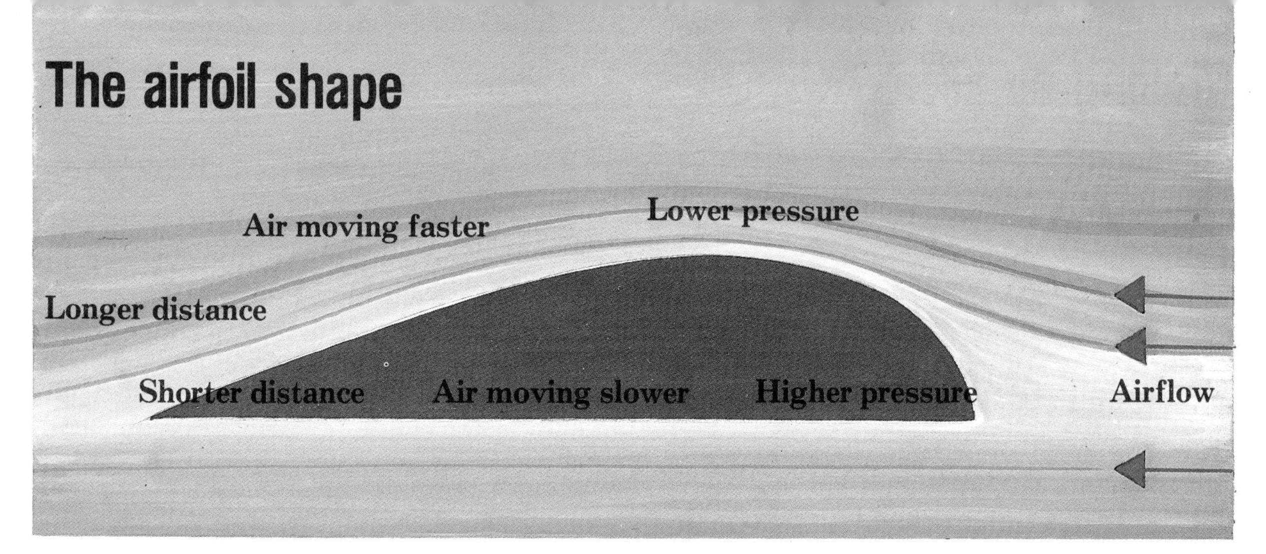

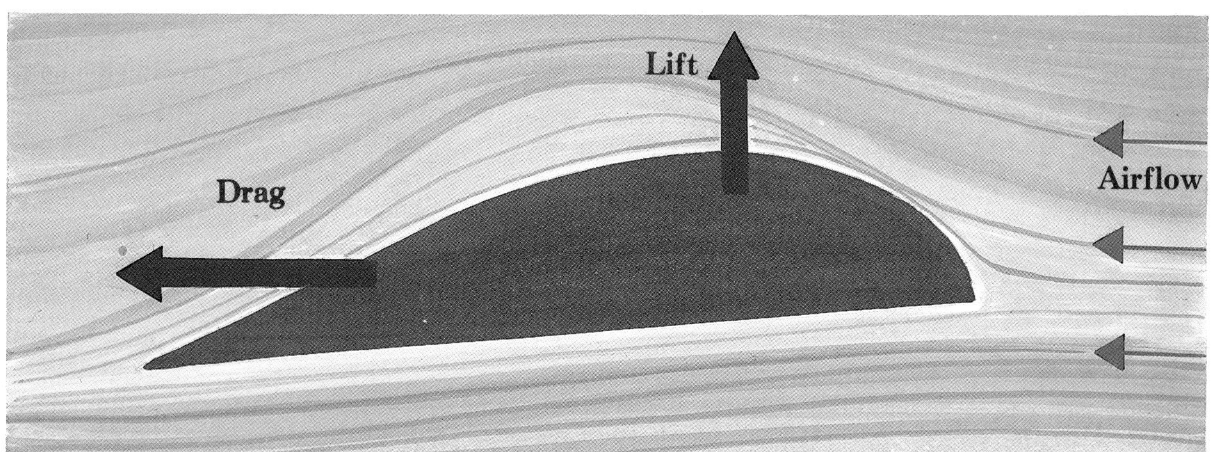

An airfoil creates lift because of the curved shape of its top surface.

This curve means that the air passing over the top of the airfoil has farther to travel than the air that flows beneath it. Because of this, the air moves faster over the top, and the air pressure above the airfoil drops. Since the pressure below the airfoil is greater than that above it, the airfoil moves upward. This is what we call lift. If you look at an aircraft, you will see that the shape of its wings is similar to the airfoil you made on page 20.

When an airfoil moves through the air, it also experiences what is called drag. This is the friction caused by the air moving over the airfoil, and it slows down the airfoil. Both the lift and the drag depend upon the shape of the airfoil. If the top surface is very curved, the airfoil creates more lift, but it also produces more drag.

Angle of attack

When an airplane is flying, the angle at which the wing meets the oncoming airflow is called the angle of attack.

Lift

Airflow

Angle of attack

If the angle of attack is increased, the amount of lift also increases.

But if the angle becomes too steep, the smooth airflow over the wing is broken, and the result is turbulence. This turbulence prevents lift.

Turbulence

If turbulence occurs, the airplane stalls, and it will fall until the pilot corrects the angle of attack. When the angle is correct, the lift force will act once again.

Stalling

Recovering

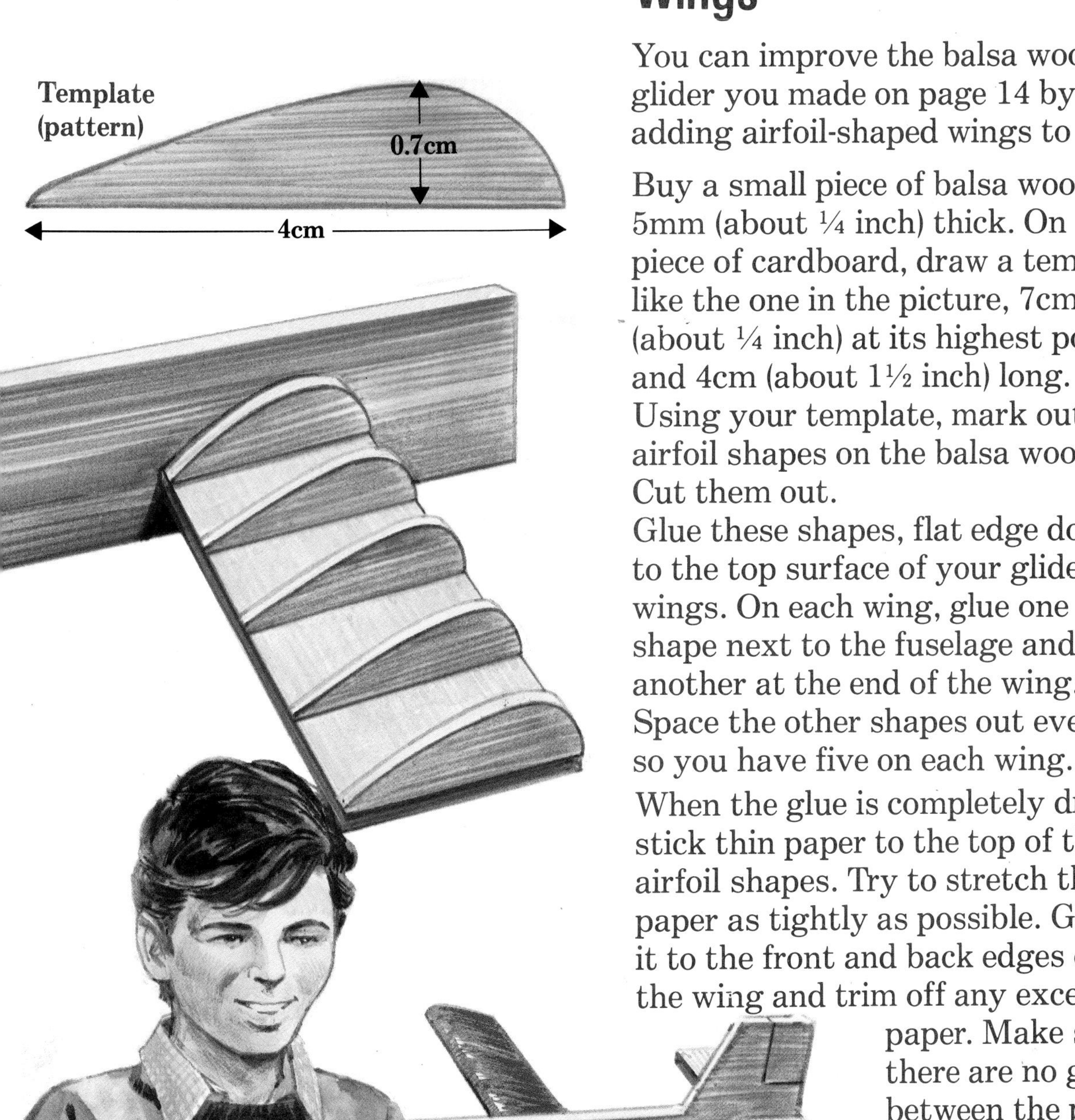

Wings

You can improve the balsa wood glider you made on page 14 by adding airfoil-shaped wings to it.

Buy a small piece of balsa wood 5mm (about ¼ inch) thick. On a piece of cardboard, draw a template like the one in the picture, 7cm (about ¼ inch) at its highest point and 4cm (about 1½ inch) long. Using your template, mark out ten airfoil shapes on the balsa wood. Cut them out.

Glue these shapes, flat edge down, to the top surface of your glider's wings. On each wing, glue one shape next to the fuselage and another at the end of the wing. Space the other shapes out evenly so you have five on each wing.

When the glue is completely dry, stick thin paper to the top of the airfoil shapes. Try to stretch the paper as tightly as possible. Glue it to the front and back edges of the wing and trim off any excess paper. Make sure there are no gaps between the paper and the wood. Allow the glue to dry and then test your glider. Does it fly better now?

Kites

When a kite is being flown, it behaves in the same way as an airfoil.

There are four forces acting upon it — lift, drag, gravity and the tension of the string.

Lift
The wind blows towards the kite and the fabric is pushed away from the frame, making an airfoil shape. Air flows more quickly over the top of the kite than beneath it. This lifts the kite upward.

Drag
The air moving over and around the kite causes drag, and this pushes the kite along.

Gravity and tension
The force of gravity pulls the kite down toward the earth. When the string is taut, the tension prevents the kite from flying away.

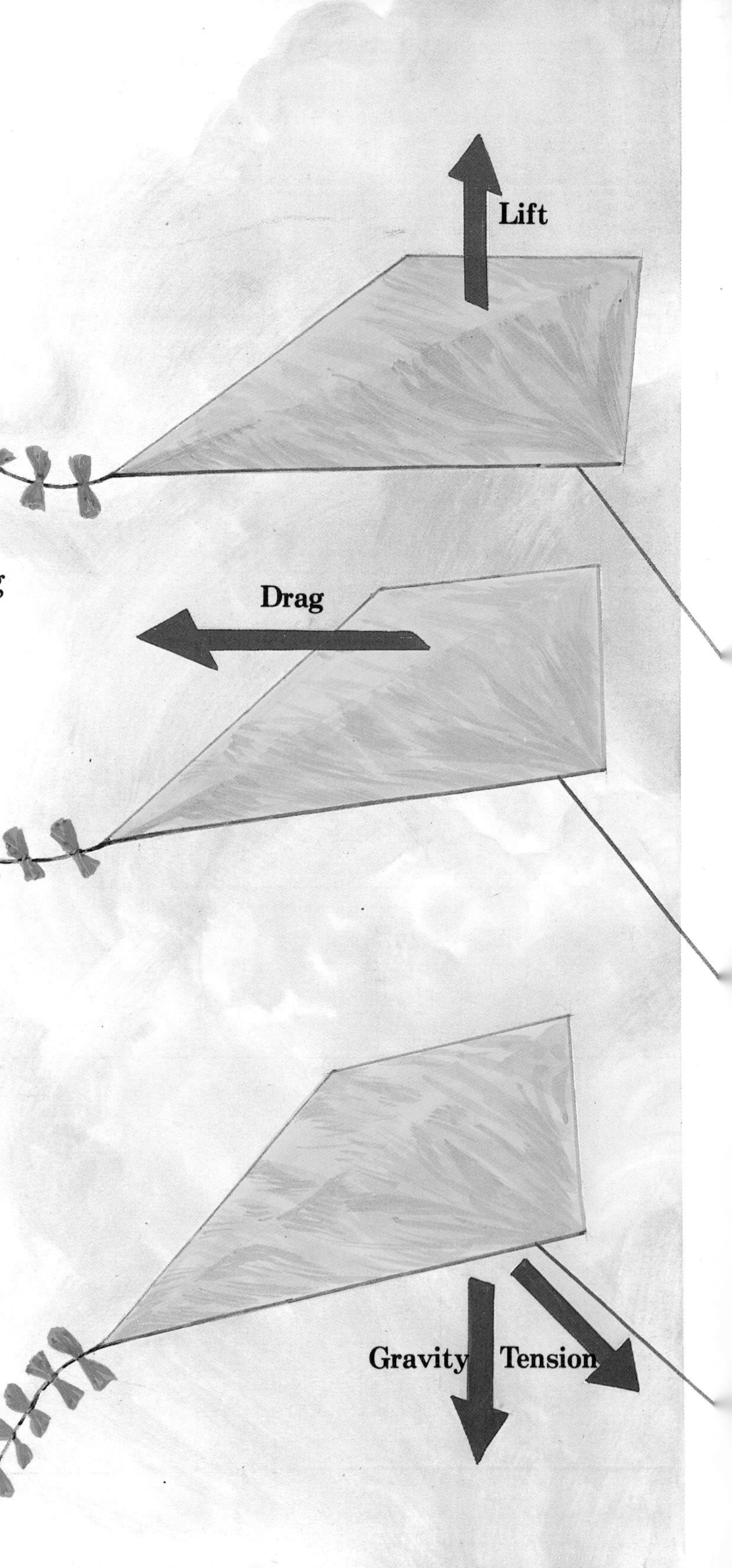

Making a kite

You can make a simple kite from two pieces of split bamboo cane or strong sticks, a sheet of thin plastic and a reel of string.
The vertical stick is about 36 inches (91cm). The horizontal stick is about 33 inches (84cm).
Join the crossbars securely with string.
Cover the frame with the plastic and secure the corners tightly with tape. Make sure the plastic is longer and wider than the frame so that you can wrap the plastic around the ends of the sticks.

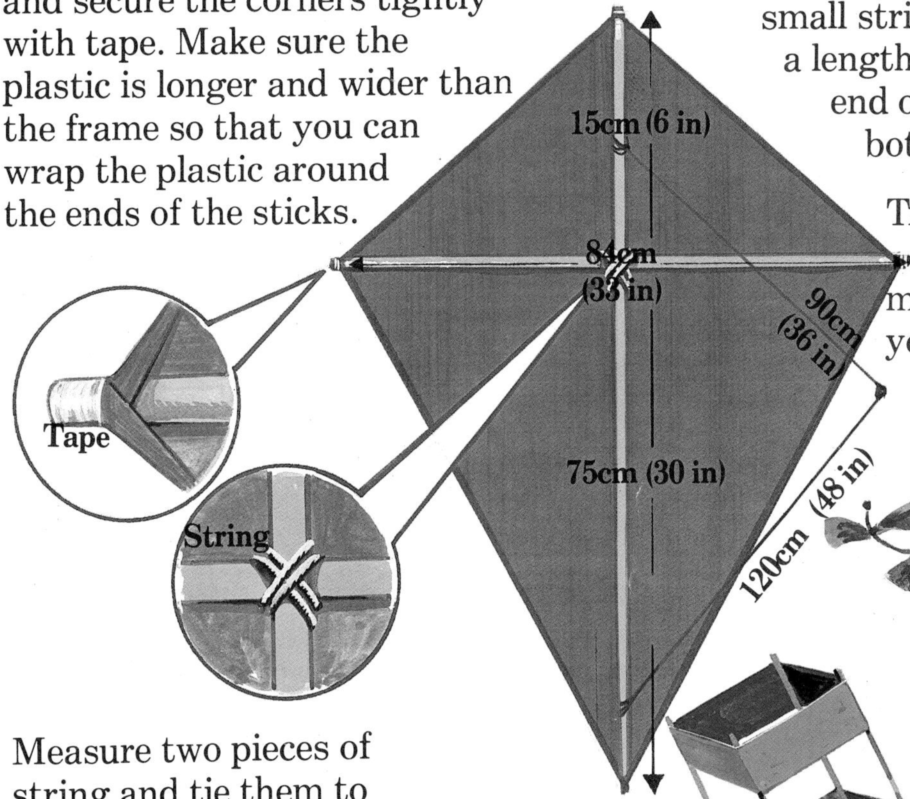

Measure two pieces of string and tie them to the upright bar as shown in the diagram. Join them together and tie on your reel of string.

To make your kite more stable, you will need to make a tail for it. The length of the tail is much more important than the weight. Make your tail about five times the length of the kite. You can make it with strips of plastic, or small strips of plastic tied to a length of string. Tie the end of your tail to the bottom of the kite.

Try these tail designs, and then make up some of your own.

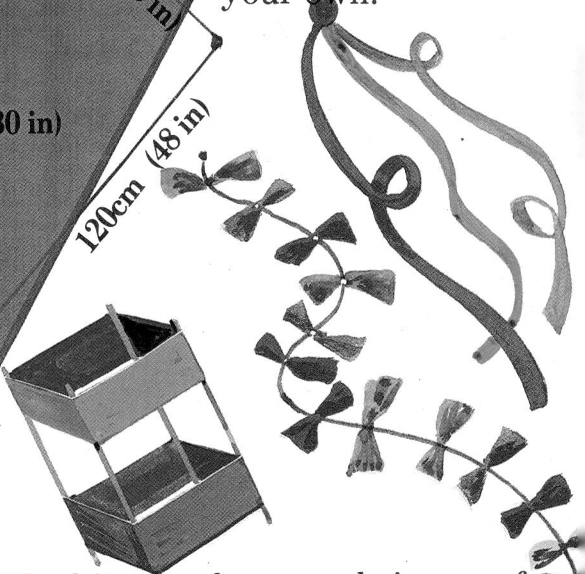

The kite you have made is one of the most simple types. Find out about other kite designs.

Powered flight

Propeller power

Buy a plastic propeller, and a piece of balsa wood about 3/8 inch wide and 8 inches long. Cut two equal lengths of strong wire and push them through the balsa-wood. Make a hook at each end of the wire. Cut the tip end from the plastic cover of a ballpoint pen. Push a short length of wire through the propeller, and then through a bead and the pen-top. Bend each end of the wire into a hook. Attach a rubber band to your machine, as shown in the picture.

Tie a length of strong thread to two supports. Make sure it is taut. Hang your machine from the thread, wind up the propeller and let go. Which way does your machine go?
Wind the propeller the other way. What happens?

Can you think of a way to adapt your propeller-powered machine so that it will power your glider?

Airplane propellers

Your propeller-powered machine moves because of the shape of its propeller. The blades of a propeller are shaped like airfoils (see page 21), and they literally pull the machine along.

Airplane propellers have the same effect. They are spun round by the airplane's engines, and they pull the plane through the air.

As we saw on page 22, an airfoil's angle of attack affects the amount of lift. In the same way, the pitch (angle) of a propeller's blades decides how far the propeller moves forward through the air in one turn.

Propellers are often called airscrews. Can you see why?

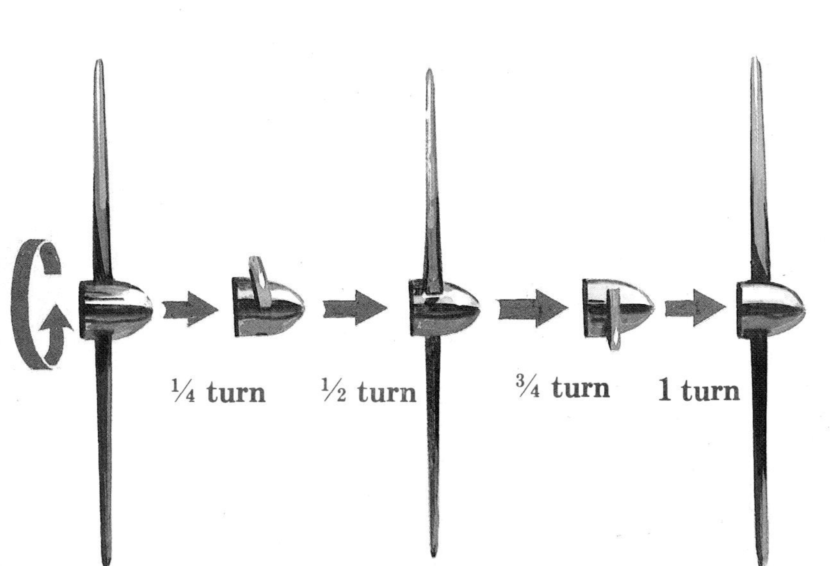

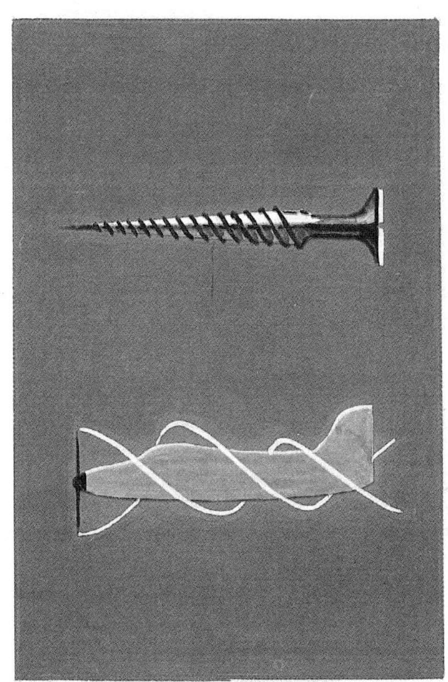

Helicopters

The rotor blades of a helicopter are rather like horizontal propellers. They are shaped like airfoils. When they spin around in the air they create lift, and the helicopter takes off.

When an airplane takes off, it first has to move along the runway to build up speed. When it is traveling fast, the wings produce enough lift for the plane to take off. A helicopter is very useful because it can take off and land without a runway.

When the rotor blades of a helicopter are spinning, they make the fuselage turn in the opposite direction. To correct this, some helicopters have a small propeller at the back. Others are fitted with two sets of rotor blades which spin in opposite directions.

Hovercraft

Rub the palms of your hands together hard. Can you feel something preventing them from rubbing together quickly? This is called friction.
A boat on the sea is slowed down by friction as the hull moves through the water. A Hovercraft is a boat that reduces this friction by riding on a cushion of air, rather than in the water. This air cushion is produced by large fans, like propellers, which blow air downward.
You can make a simple Hovercraft from a polystyrene ceiling tile. Cut your tile into the shape of a boat.
Glue two of the corners of the tile onto the back of the boat shape.
Cut out a 6cm (2¼ inch) hole from the center of the shape.
Use a hairdryer to blow into the hole. Your Hovercraft should lift.
To get it moving, give it a slight push.

Jet propulsion

Blow up a balloon and let it go.

The balloon flies through the air but its flight is uncontrolled. The air inside the balloon is forced out and this causes the balloon to move. This is called thrust.

Pass a length of strong thread through a plastic straw. Fasten the ends of the thread to two supports and make sure it is taut. Push two short lengths of straw into the neck of a balloon. Hold them in place with a rubber band.
Blow up the balloon and twist the neck to keep the air in.

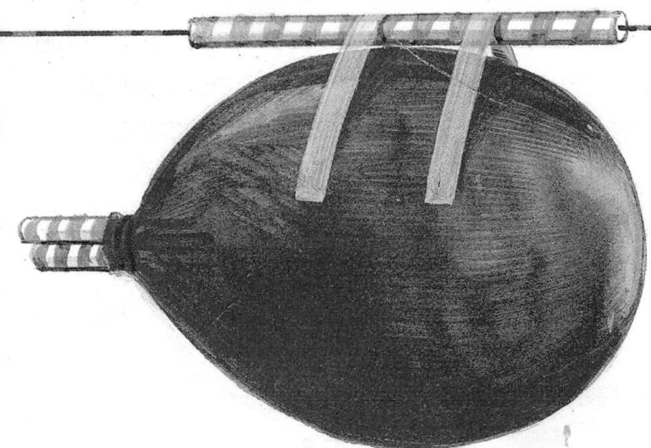

Fix the balloon to the long straw with sticky tape.
Untwist the neck of the balloon and let it go.
How far does it travel?
Blow up your balloon with different amounts of air.
Does this alter the distance it will travel?

Jet engines

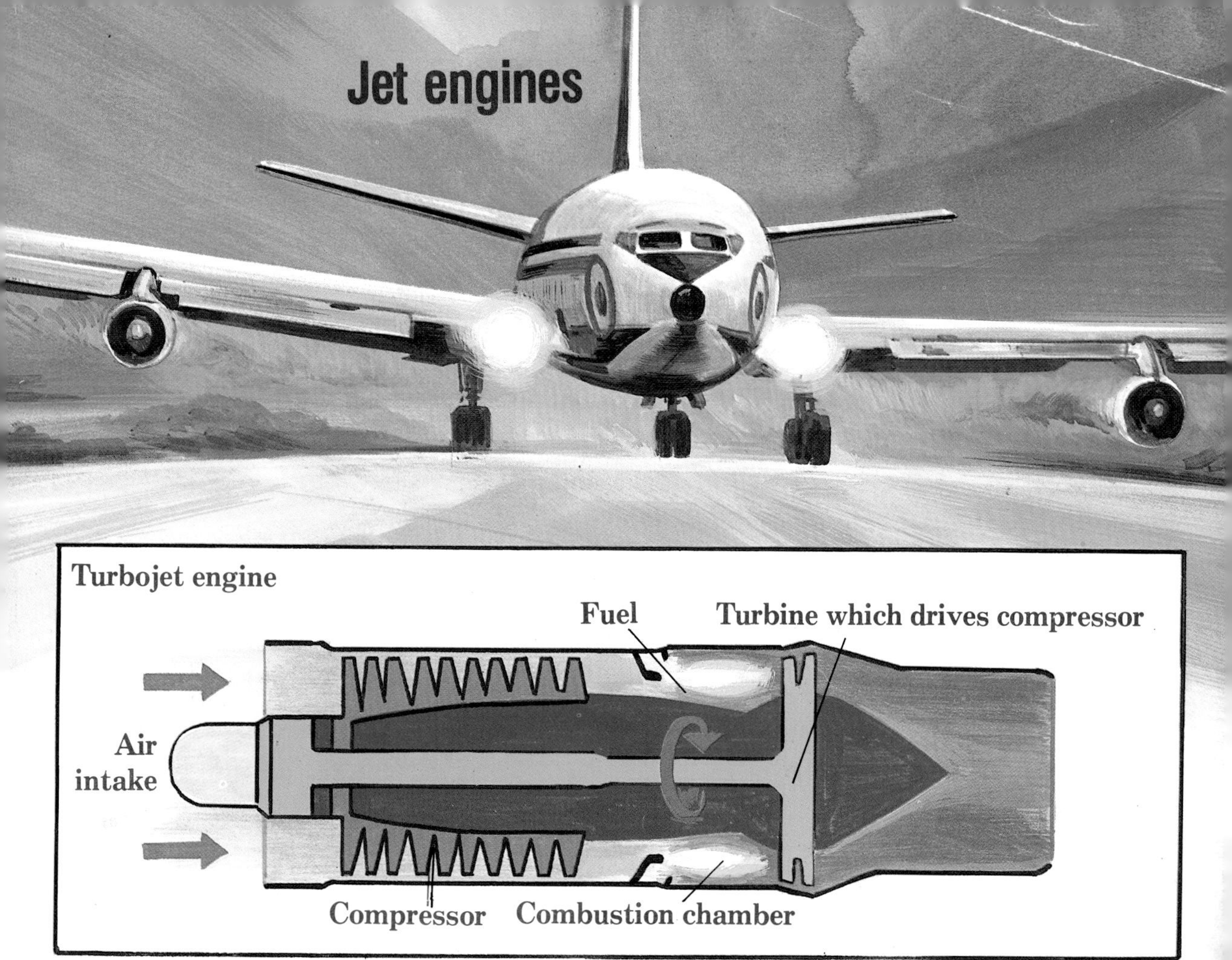

A jet engine works in a similar way to your balloon "jet" on page 30. Air is taken in at the front of the engine and forced out at the back to give an airplane forward thrust.

The most widely used type of jet engine today is called a turbojet. It has three main parts— a compressor, combustion chambers, and a turbine.

Air is drawn in at the front and is then compressed (squeezed). In the combustion chamber, fuel is mixed with the compressed air and the mixture burns. The hot gases expand and force their way out of the combustion chamber. They hit the blades of the turbine causing it to spin, rather like a large fan. The spinning turbine is connected to the compressor, and this draws more air into the front of the engine. The hot gases are then pushed out of the back of the engine at high speed, producing thrust to push the airplane through the air.

Rockets

A jet can operate only in the earth's atmosphere, because it needs a constant supply of oxygen (air) entering the front of the engine.
A rocket works in a similar way to a jet engine, except that it carries both fuel and oxygen with it. This means that rockets can fly in space, where there is no oxygen.

Inside a rocket, fuel and oxygen are burned fiercely in a combustion chamber. This produces very hot gases under high pressure. These gases force their way out of a nozzle at the back, pushing the rocket forward.